Recovering Silver from E-Waste Guide for Beginners

Roggie Clark

Continue Learning At Our Website: Roggieclarkbooks.com

Continue exploring educational resources, tools, books, workbooks, and learning materials designed to help readers continue building knowledge and practical skills. Visit: www.roggieclarkbooks.com

Inside you'll find:

- Educational worksheets
- Beginner-friendly guides
- Printable activities
- Companion materials for books and courses

A growing library of educational content for readers, collectors, learners, parents, teachers, and lifelong students.

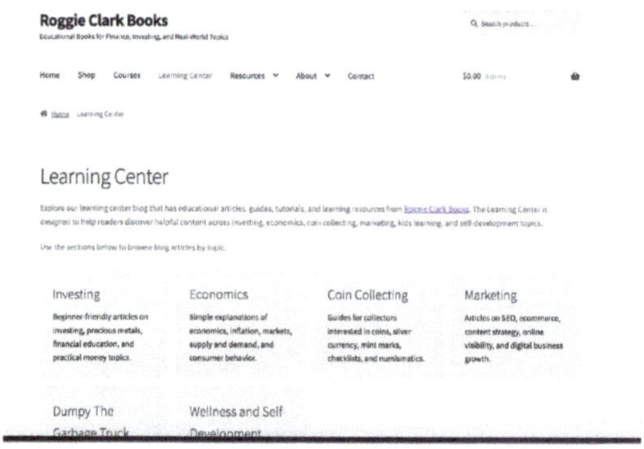

Learn Macro and Micro Economics, With This One Book

Whether you're a student, a professional, or simply someone eager to understand the world around you, this book is your perfect introduction to the fundamentals of micro and macroeconomics.

https://www.amazon.com/Macro-Micro-Economics-Workbook-Beginners-ebook/dp/B0FVPMFX6B

The Strategic Guide to U.S. Mint Gold and Silver Coin Investing

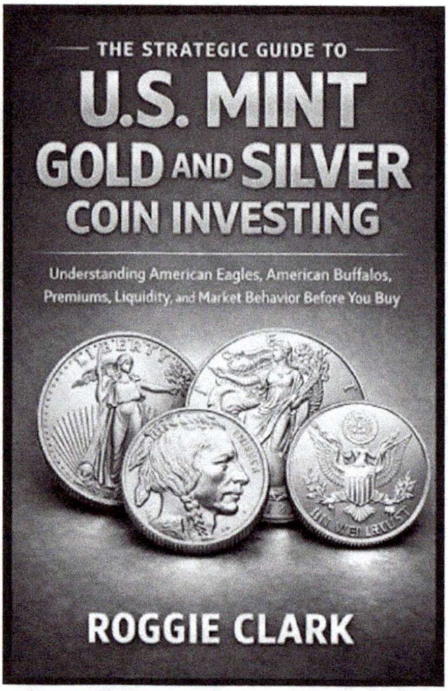

Rather than offering a general introduction to precious metals, this guide examines how American Gold Eagles, American Gold Buffalos, and Silver American Eagles behave in real investor markets.

https://www.amazon.com/Strategic-Guide-U-S-Silver-Investing/dp/B0GQKJF9K6

Copyright

Title: Recovering Silver from E-Waste Guide for Beginners
Author: Roggie Clark

Cover Design: Roggie Clark
Editing: Renee' Clark

Printed in the United States of America

Legal and Safety Disclaimer

The information presented in this book is provided solely for general educational and informational purposes. It is intended to help readers better understand silver-bearing electronic materials, electronic waste (e-waste), precious metals recycling, and responsible sourcing practices. It should not be interpreted as legal, financial, tax, environmental, engineering, electrical, chemical, recycling, or professional business advice.

The author and publisher make every effort to present accurate and current information. However, laws, regulations, environmental requirements, hazardous waste disposal rules, chemical handling standards, workplace safety requirements, recycling policies, auction procedures, and buyer acceptance standards vary by country, state, province, municipality, and organization, and they may change over time. Readers are solely responsible for researching and complying with all applicable laws, regulations, permits, and safety requirements within their jurisdiction.

This book promotes only **legal and ethical methods** of sourcing, collecting, identifying, recovering, transporting, and selling silver-bearing materials. Nothing contained in this publication should be interpreted as encouraging theft, trespassing, unauthorized removal of property, unlawful collection of electronic waste, improper disposal of hazardous materials, environmental violations, or any other illegal or unethical activity. Always obtain permission before collecting, purchasing, dismantling, transporting, or processing materials that belong to another individual, business, or organization.

This book discusses silver recovery methods solely to provide readers with a general understanding of how the precious metals recycling industry operates. References to mechanical separation, chemical recovery, acid peroxide, aqua regia, electrolytic refining, or other refining methods are provided for educational purposes only. They are **not** intended as step-by-step instructions or recommendations for performing chemical refining.

Chemical refining involves potentially hazardous substances that can produce toxic fumes, chemical burns, fires, environmental contamination, and other serious risks if handled improperly. Safe refining requires specialized knowledge, appropriate laboratory equipment, proper ventilation, suitable personal protective equipment, hazardous waste management procedures, and compliance with all applicable environmental and workplace safety regulations. Readers should not attempt chemical refining without appropriate training, experience, facilities, and legal authorization.

Many electronic devices also contain hazardous materials unrelated to silver recovery, including batteries, capacitors, lead-containing solder, mercury-containing components, sharp metal edges, glass, chemicals, and other materials that require careful handling and proper disposal. Always follow manufacturer guidance, applicable safety regulations, and local hazardous waste disposal requirements when handling electronic equipment.

References to refiners, recycling companies, businesses, auctions, organizations, manufacturers, government agencies, products, services, or equipment are provided solely for educational purposes and do not constitute endorsements or recommendations. Pricing, grading standards, purchasing policies, refining methods, and acceptance requirements differ among organizations and may change without notice.

Any discussion of potential profits, resale values, recovered metal yields, business opportunities, or financial returns is illustrative only and should not be interpreted as a guarantee of future results. Actual outcomes vary based on material quality, market conditions, precious metal prices, sourcing methods, operating costs, refining efficiency, experience, and numerous other factors beyond the control of the author and publisher.

By using the information contained in this book, you acknowledge that you assume full responsibility for your own decisions and actions. The author and publisher disclaim all liability for any loss, injury, illness, property damage, environmental damage, financial loss, legal consequences, regulatory penalties, or other claims arising directly or indirectly from the use or misuse of the information presented in this publication.

Image Credits and Public Domain Notice

All images, illustrations, graphics, photographs, visual references, and related materials included in this publication are believed to be in the public domain, used under applicable licenses or permissions, created by the author or publisher, or obtained from sources that clearly identify them as available for lawful use.

The author and publisher have made a good faith effort to use only visual materials that are lawfully available for use and distribution. Where applicable, images and visual materials may have been resized, cropped, edited, enhanced, stylized, modified, reformatted, or arranged for educational, informational, design, formatting, or presentation purposes.

This publication is intended for educational, informational, and illustrative purposes only. The inclusion of any image, artwork, photograph, visual reference, symbol, historical material, design, or graphic element is not intended to claim ownership over public domain materials or any third-party intellectual property rights that remain with their respective owners.

If you believe any image or visual material included in this publication has been used in error, please contact the author or publisher so the matter may be reviewed and corrected in future editions or releases if necessary.

Any factual information, specifications, historical references, values, descriptions, statistics, or related content accompanying visual materials are provided for general informational purposes only and may change over time. Readers should conduct their own independent research and consult appropriate sources or qualified professionals when making decisions based on information presented in this publication.

This publication is independently produced and is not affiliated with, endorsed by, sponsored by, or associated with any company, organization, government agency, institution, trademark holder, or third party unless specifically stated otherwise.

Table of Contents

Introduction

Silver has fascinated people for thousands of years, but its importance extends far beyond coins, jewelry, and collectibles. Today, silver plays a critical role in modern technology because it is one of the most electrically and thermally conductive metals on Earth. From computers and industrial equipment to electrical relays, switches, connectors, motors, and telecommunications systems, silver is used in countless products that power everyday life. As these products reach the end of their useful lives, many become part of the growing stream of electronic waste, creating opportunities for responsible recyclers who understand where silver is found and how to identify it. While most people see discarded electronics, experienced recyclers recognize the valuable silver-bearing components hidden inside them.

Why Silver Recovery Has Become So Popular

As interest in electronic recycling continues to grow, more people are discovering that electronic devices contain much more than copper wiring and circuit boards. Silver is commonly used in electrical contacts, relays, switches, connectors, keyboards, industrial controls, and many other components where excellent electrical conductivity and resistance to wear are essential. Although the amount of silver contained in any single component is usually small, these materials can accumulate into valuable quantities over time when sourced responsibly.

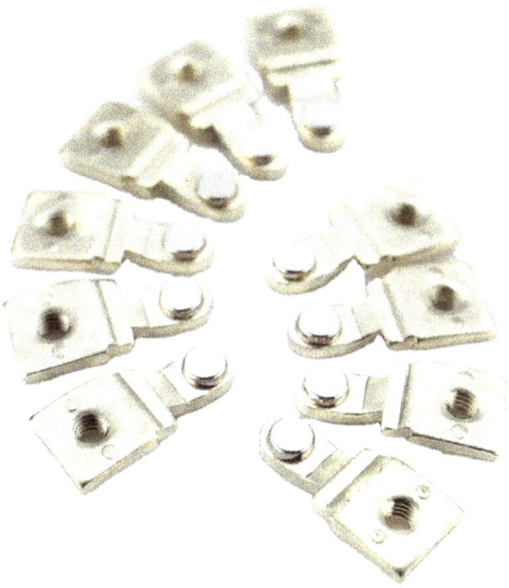

The popularity of online videos featuring precious metal recycling has introduced millions of people to the world of silver recovery. Viewers often watch recyclers sorting relay contacts, dismantling industrial equipment, identifying silver-plated components, or discussing refining methods, and naturally begin asking where these materials come from and whether they can recover silver themselves. While these videos can be educational, they sometimes leave out the larger picture of responsible sourcing, realistic expectations, and the knowledge needed to participate safely in the recycling industry.

One of the most important lessons experienced recyclers learn is that recovering silver is only one part of the process. Successfully identifying silver-bearing materials, understanding their value, sourcing them legally, and deciding whether to sell them or send them to professional refiners are often far more important than attempting to refine the metal yourself.

How This Book Will Help You

Beginning in the precious metals recycling industry can seem overwhelming. There are many different types of silver-bearing materials, numerous testing methods, changing market prices, and a wide range of opinions about recovery techniques. Some sources focus entirely on chemical refining, while others only discuss collecting electronic scrap without explaining why certain components are valuable.

This guide is designed to provide a balanced introduction.

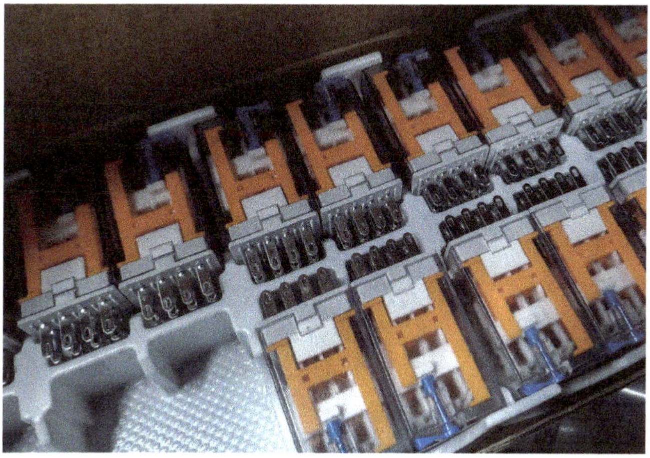

You'll learn where silver is commonly found, why manufacturers use it, how to recognize silver-bearing components, and where these materials can be legally sourced. You'll also explore how buyers evaluate silver-bearing scrap, the factors that influence its value, and the practical differences between selling materials to professional refiners and attempting to recover silver yourself.

Because many readers are curious about chemical recovery, this book also includes a high-level overview of the most common recovery methods used within the industry. Rather than providing laboratory procedures, the goal is to help you understand how these processes work conceptually, why they require specialized knowledge, and why many recyclers choose to sell silver-bearing materials to professional refiners instead.

Whether your goal is earning supplemental income, expanding your recycling knowledge, building an ethical side business, or simply understanding how precious metals move through the recycling economy, this guide will provide the foundation needed to begin confidently.

Ethical, Legal, and Safety First

Responsible silver recovery begins with ethics, legality, and safety. Every silver-bearing component discussed throughout this book should be obtained through lawful means, including electronics recycling programs, public auctions, business liquidations, surplus sales, industrial equipment upgrades, and other authorized sources. Respecting property rights and obtaining permission before collecting or purchasing materials is essential for building a successful long-term recycling business.

SAFETY
FIRST
ISO
14001

It is equally important to recognize that some discussions of silver recovery involve hazardous chemicals and industrial refining techniques. Professional refiners use specialized equipment, controlled environments, protective gear, and carefully managed waste handling procedures to safely recover precious metals. These processes require significant training, careful planning, and compliance with environmental regulations. Throughout this guide, chemical recovery is presented only to help readers understand how the industry works, not as encouragement to perform hazardous refining without the appropriate expertise and facilities.

Approaching silver recovery with integrity benefits everyone involved. Suppliers appreciate responsible recyclers, buyers value accurately represented materials, and communities benefit when valuable resources are recovered while electronic waste is managed responsibly. Building your knowledge on this foundation will serve you far better than chasing shortcuts or unrealistic expectations.

Last Thoughts

Silver recovery is about much more than extracting a precious metal from electronic scrap. It is about understanding where silver is used, recognizing valuable opportunities, making informed decisions, and participating responsibly in an industry that helps conserve valuable natural resources. By learning why silver recovery has become so popular, understanding what this book is designed to teach, and committing yourself to ethical and safe practices, you'll be well prepared for the chapters ahead. In the next chapter, you'll begin exploring the fundamentals of silver itself, including why it is used in electronics, where it is commonly found, and how understanding silver-bearing materials will help you make better decisions as your knowledge continues to grow.

Chapter 1: Understanding Silver and Silver-Bearing Scrap

Before you begin searching for silver-bearing materials, it's important to understand why silver is used in electronics and why certain components are more valuable than others. Many beginners assume silver is found only in jewelry, coins, or decorative items, but modern industry depends on silver in countless electrical and electronic applications. Learning where silver is commonly used, how manufacturers apply it, and why it is often combined with other metals will help you recognize worthwhile opportunities while developing a stronger understanding of the precious metals recycling industry.

Why Silver Is Used in Electronics

Silver is one of the most remarkable industrial metals because it possesses the highest electrical conductivity and thermal conductivity of any naturally occurring metal. These properties allow electricity and heat to move through silver with exceptional efficiency, making it ideal for applications where reliable electrical performance is critical. Manufacturers often choose silver for components that must carry electrical current repeatedly while resisting wear and maintaining dependable performance over many years.

Silver contact
Copper coils

Pure copper
Silver alloy
electroshock

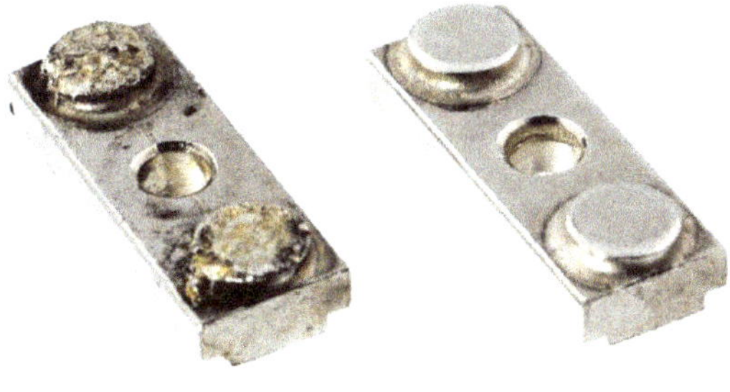

Unlike copper, which is commonly used for wiring because of its lower cost, silver is generally reserved for specialized applications where superior conductivity or resistance to electrical arcing provides an important advantage. Electrical relays, switches, contactors, circuit breakers, industrial controls, and many types of electronic connectors often contain small silver-bearing contact surfaces that repeatedly open and close electrical circuits. These components experience thousands or even millions of switching cycles during their service life, making silver's performance characteristics especially valuable.

Silver also resists corrosion better than many other conductive materials, although it does develop a thin layer of tarnish over time. This tarnish differs from rust and often appears as a gray, bluish, or slightly iridescent discoloration. While tarnish can affect certain high-frequency electrical applications, the metal underneath generally remains intact, allowing silver-bearing components to continue providing value even after years of service.

Understanding why manufacturers invest in silver helps explain why recyclers continue searching for these components today. Although each individual contact or connector may contain only a small amount of silver, millions of these components enter the recycling stream every year as equipment is replaced or upgraded.

Common Sources of Silver-Bearing Scrap

Silver is found in far more places than many beginners expect. Rather than existing in large solid pieces, it is usually present in small electrical contact points, plated surfaces, specialized connectors, and industrial components designed for demanding electrical applications.

1K63
1K62
30K3
1

Electrical relays remain one of the most recognizable sources of silver-bearing contacts. Inside many relays are small contact buttons that repeatedly complete electrical circuits whenever the relay activates. Similar contact points are commonly found inside industrial switches, circuit breakers, motor starters, and electrical contactors used in commercial and industrial facilities. Because these components interrupt electrical current thousands of times during their service life, manufacturers frequently use silver alloys to improve durability and reduce electrical arcing.

Silver may also be found in certain connectors, ribbon cable terminations, industrial control systems, telecommunications equipment, electric motors, and specialized electronic devices. Older computer keyboards have historically used silver-bearing conductive layers, while some industrial equipment contains silver-plated electrical components designed for long-term reliability. Not every electronic device contains significant quantities of silver, but learning which categories of equipment commonly use silver allows you to focus your efforts more effectively.

As your experience grows, you'll begin recognizing these components almost immediately. Instead of viewing electronic equipment as a collection of unrelated parts, you'll start seeing individual opportunities hidden within larger assemblies. This ability develops through observation, repeated exposure, and continuous learning rather than memorizing every possible product.

Understanding Silver Contacts, Plating, and Alloys

One of the biggest surprises for beginners is discovering that many silver-bearing components are not made from pure silver. Manufacturers frequently combine silver with other metals to improve strength, durability, resistance to wear, or performance under repeated electrical switching.

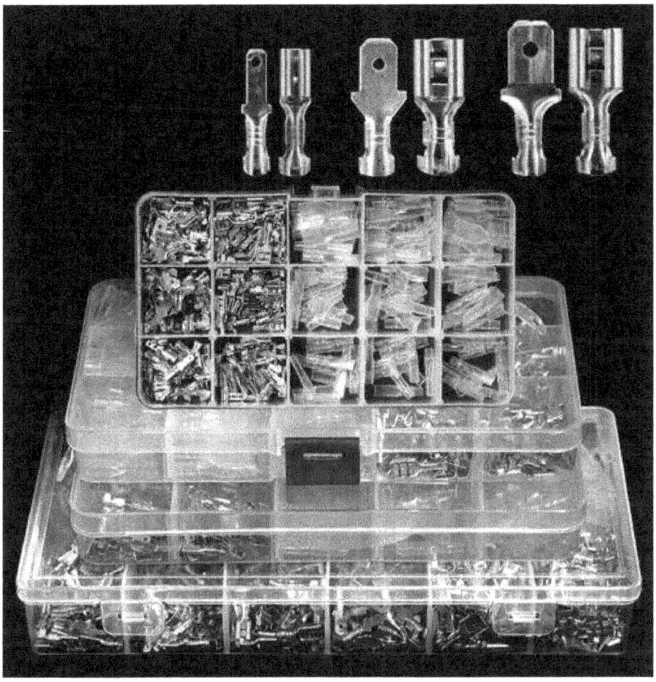

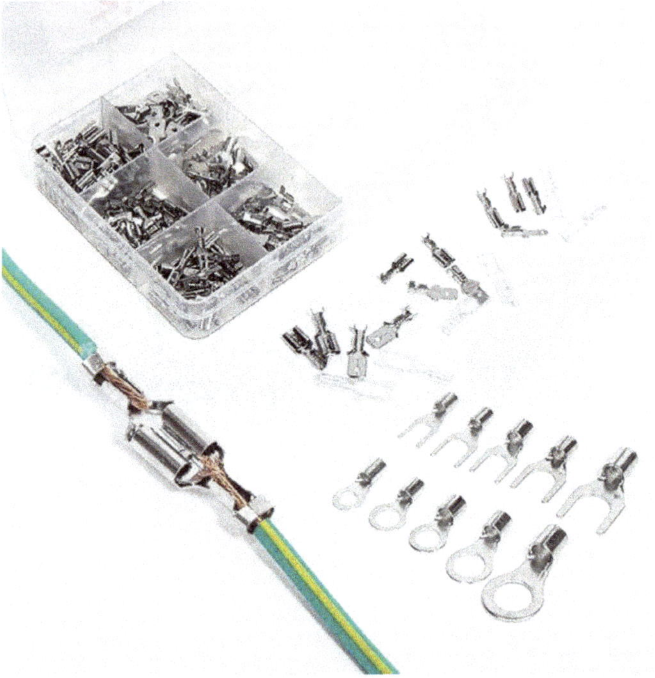

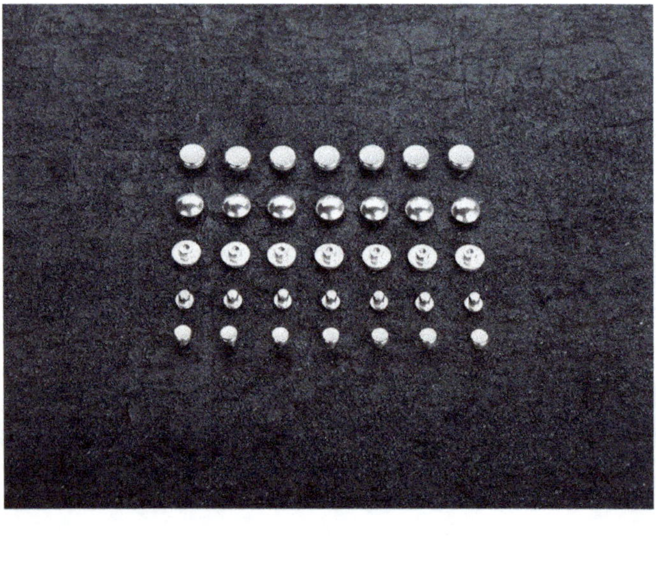

Many electrical contacts are manufactured using **silver alloys** rather than pure silver. Depending on their intended application, manufacturers may combine silver with metals such as tungsten, cadmium, nickel, or other materials that improve resistance to heat, electrical arcing, and mechanical wear. Likewise, many connectors and terminals are only **silver-plated**, meaning a thin layer of silver has been applied over another base metal rather than the entire component being made from silver.

This distinction is important because appearance alone rarely reveals how much silver is actually present. A component that looks completely silver-colored may contain only a thin plated surface, while another small contact point may contain a much higher percentage of recoverable silver despite its modest size. Understanding these differences helps explain why professional buyers carefully sort and evaluate materials rather than relying solely on visual inspection.

Testing methods can also help identify silver-bearing materials. Visual characteristics, manufacturer markings, and commercially available silver testing products may provide useful information when evaluating certain items. However, no single method is perfect, and experienced recyclers often combine multiple observations before drawing conclusions. Learning to recognize common silver-bearing components through experience is usually more valuable than relying on appearance alone.

As you continue through this book, you'll build on this knowledge by learning where to legally source these materials, how buyers evaluate them, and the practical options available for selling or recovering their silver content responsibly.

Last Thoughts

Understanding silver and the materials that contain it is the foundation of successful precious metals recycling. By learning why silver is used in electronics, recognizing common silver-bearing components, and understanding the differences between silver contacts, plating, and alloys, you'll be better prepared to identify valuable opportunities with confidence. In the next chapter, you'll explore the many legal and ethical places where silver-bearing materials can be found, from household electronics and industrial equipment to auctions, business liquidations, and other legitimate sourcing opportunities.

Chapter 2: Finding Silver-Bearing Materials Legally

Once you understand why silver is used in electronics and how manufacturers incorporate it into electrical components, the next step is learning where these materials come from. This is one of the most common questions beginners ask after seeing recyclers collect bins of silver contacts, relays, and industrial electrical equipment. The reality is that experienced recyclers do not rely on luck. They understand where silver-bearing equipment naturally reaches the end of its service life and develop relationships with people and organizations that regularly replace or retire these products. By focusing on ethical and legal sourcing methods, you can steadily build your knowledge and inventory while contributing to responsible recycling practices.

Residential Electronics and Household E-Waste

Many people are surprised to learn that opportunities to find silver-bearing materials often begin at home. Computers, printers, televisions, audio equipment, power tools, kitchen appliances, networking devices, and older electronics all contain electrical switches, connectors, relays, or control boards that may include small amounts of silver.

Friends, family members, neighbors, and coworkers frequently store outdated electronics in garages, closets, or basements simply because they are unsure how to dispose of them. Letting people know that you responsibly recycle electronic equipment can create a steady source of materials while helping others reduce clutter. Over time, these small opportunities often become one of the easiest ways for beginners to build experience.

Household appliances also deserve attention. Microwave ovens, washing machines, dryers, furnaces, air conditioners, and other large appliances often contain switches, relays, timers, and electrical controls that use silver-bearing contacts. Although not every appliance contains significant quantities of silver, these components become valuable when viewed as part of a much larger recycling effort rather than isolated finds.

Older electronic equipment can sometimes be particularly interesting because manufacturers historically relied more heavily on mechanical relays and larger electrical contacts than many modern devices. As technology evolved, component designs changed, but many older systems still contain silver-bearing parts that eventually enter the recycling stream.

Community electronics recycling events also provide an excellent opportunity to observe the types of equipment commonly entering the waste stream. While collection policies vary by location, these programs demonstrate just how much electronic equipment reaches the end of its useful life each year.

Industrial, Commercial, and Electrical Sources

While household electronics provide valuable learning opportunities, many experienced recyclers eventually expand into commercial and industrial sources where larger quantities of silver-bearing components are often found.

Manufacturing facilities, office buildings, hospitals, schools, utility companies, warehouses, and industrial plants regularly replace electrical equipment during maintenance, modernization projects, and facility upgrades. Electrical contractors removing outdated control systems often encounter relays, motor starters, contactors, disconnect switches, circuit breakers, and other devices that contain silver-bearing electrical contacts. These components are specifically designed to withstand repeated electrical switching, making silver an ideal material for their contact surfaces.

Computer repair businesses, industrial maintenance companies, and electrical service contractors may also generate electronic scrap as part of their routine operations. Developing professional relationships with these businesses can create consistent opportunities while helping them responsibly recycle unwanted equipment.

Telecommunications equipment represents another important category. Telephone systems, networking infrastructure, communication hardware, and industrial control systems frequently contain relays, connectors, switching equipment, and specialized circuit boards that may include silver-bearing components. As these systems are upgraded or replaced, they become part of the broader electronic recycling economy.

The most successful recyclers understand that reliable sourcing is built on trust. By communicating professionally, respecting property rights, arriving when promised, and handling materials responsibly, you create relationships that often produce opportunities long after a single transaction has ended.

Auctions, Liquidations, and Other Legal Opportunities

Some of the largest collections of silver-bearing equipment become available through organized surplus sales rather than individual electronic devices. Public auctions, business liquidations, government surplus programs, and industrial equipment sales regularly offer electrical equipment that may contain valuable silver-bearing components.

WHALLON

WESTON
849A

SQD 22KA
QOB350
QOB360
SQD
QOB GFI
10KA
22KA
SIEMENS
B240 / B250
SIEMENS
B220
SIEMENS
B330
SIEMENS

Businesses periodically replace manufacturing equipment, electrical distribution systems, laboratory instruments, telecommunications hardware, and office technology. Rather than disposing of these materials, many organizations sell them through public auctions or liquidation firms. Learning how these sales operate allows recyclers to purchase equipment legally while gaining access to categories of electronic scrap that are rarely available through household sources.

Government agencies, schools, universities, hospitals, and utility companies also retire equipment on a regular basis. Public surplus auctions may include electrical panels, industrial controls, laboratory electronics, communications equipment, and other specialized devices containing silver-bearing components.

Estate sales, warehouse cleanouts, storage unit auctions, and business closures may also present worthwhile opportunities. However, experienced buyers remain selective. Instead of purchasing every available item, they evaluate equipment based on the likelihood of containing valuable components, the amount of work required, transportation costs, and potential resale or recycling value.

Over time, experienced recyclers spend less effort searching randomly because they have developed dependable sourcing channels. Building those relationships requires patience, professionalism, and consistency, but the long-term benefits often far outweigh relying on occasional discoveries.

Last Thoughts

Finding silver-bearing materials legally begins with understanding where electrical and electronic equipment naturally reaches the end of its useful life. Homes, commercial buildings, industrial facilities, electrical contractors, public auctions, surplus sales, and business liquidations all provide legitimate opportunities for responsible recyclers. By focusing on ethical sourcing and developing long-term relationships, you'll create a strong foundation for future success. In the next chapter, you'll learn how to identify silver-bearing components more confidently, organize your materials efficiently, and use the basic tools that help experienced recyclers evaluate and prepare silver-bearing scrap.

Chapter 3: Identifying, Collecting, and Preparing Silver-Bearing Scrap

Once you've learned where silver-bearing materials are commonly found, the next step is learning how to recognize them with greater confidence. This is often where beginners make the greatest progress because experience gradually replaces guesswork. While no one can identify every valuable component at a glance, understanding the characteristics of silver-bearing materials, organizing your findings properly, and using a few basic tools will help you make better decisions over time. This chapter focuses on building practical skills that improve both efficiency and confidence while preparing your materials for sale or further evaluation.

Identifying Silver-Bearing Components

The ability to identify silver-bearing components is developed through observation rather than memorization. Although silver may be found in many different types of equipment, it most often appears where reliable electrical switching or electrical contact is required. Components that repeatedly open and close electrical circuits are among the most common places to begin looking.

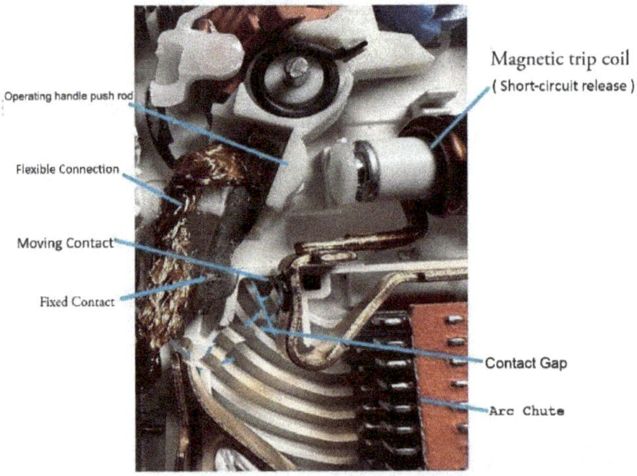
Magnetic trip coil
(Short-circuit release)
Operating handle push rod
Flexible Connection
Moving Contact
Fixed Contact
Contact Gap
Arc Chute

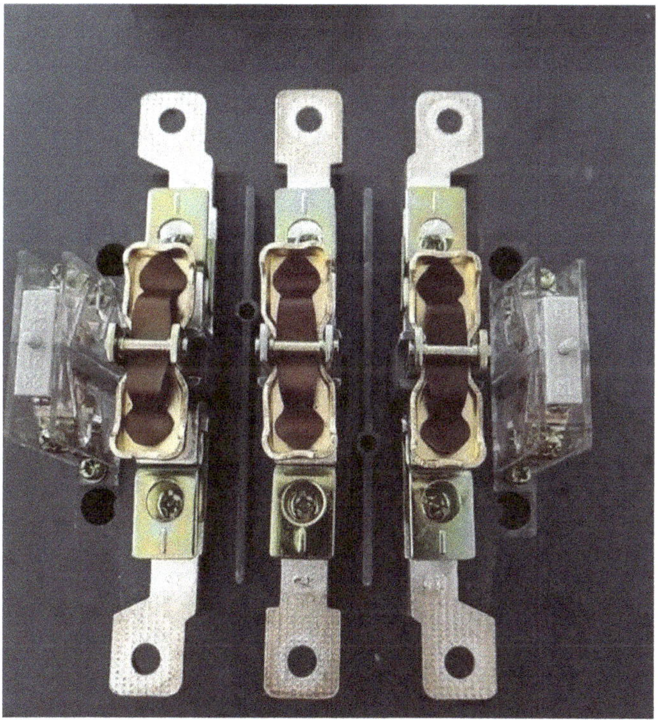

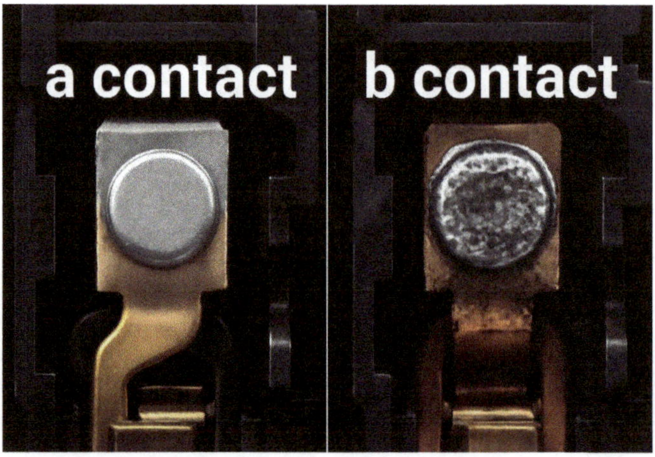

Electrical relays, contactors, circuit breakers, disconnect switches, motor starters, and industrial control equipment frequently contain small contact buttons manufactured from silver alloys. These contacts are designed to withstand repeated electrical arcing while maintaining reliable conductivity over many years of operation. Although the contacts themselves may appear small, they are often the most valuable part of the device from a precious metals recycling perspective.

Silver-bearing materials may also appear on certain electrical connectors, ribbon cable terminations, industrial switches, telecommunications equipment, and specialized electronic devices. Some older keyboards and electrical assemblies also contain silver-based conductive components, although the amount of recoverable silver varies considerably depending on the design and manufacturer.

Visual clues can also be helpful. Silver often develops a distinctive tarnish over time that differs from ordinary corrosion. Instead of reddish rust or green oxidation commonly associated with iron or copper, tarnished silver may develop gray, bluish, or slightly iridescent tones. While appearance alone should never be considered proof that a component contains silver, recognizing these characteristics helps narrow your search.

As your experience grows, you'll begin identifying likely silver-bearing components much more quickly. Instead of disassembling every electronic device completely, you'll learn which areas deserve closer inspection and which components are less likely to contain valuable precious metals.

Sorting, Testing, and Organizing Your Scrap

After collecting silver-bearing materials, careful organization becomes one of the most valuable habits you can develop. Proper sorting makes identification easier, simplifies conversations with buyers, and prevents higher-value materials from becoming mixed with ordinary electronic scrap.

B
C
E

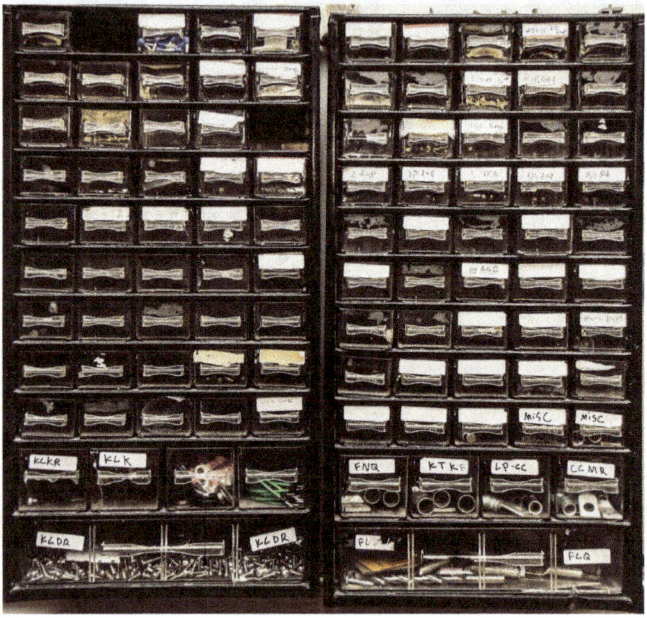
KLKR
KLK
FNQ
LP-CC
CCMR
MiSC
MiSC
KLDR
KLDR
FLQ

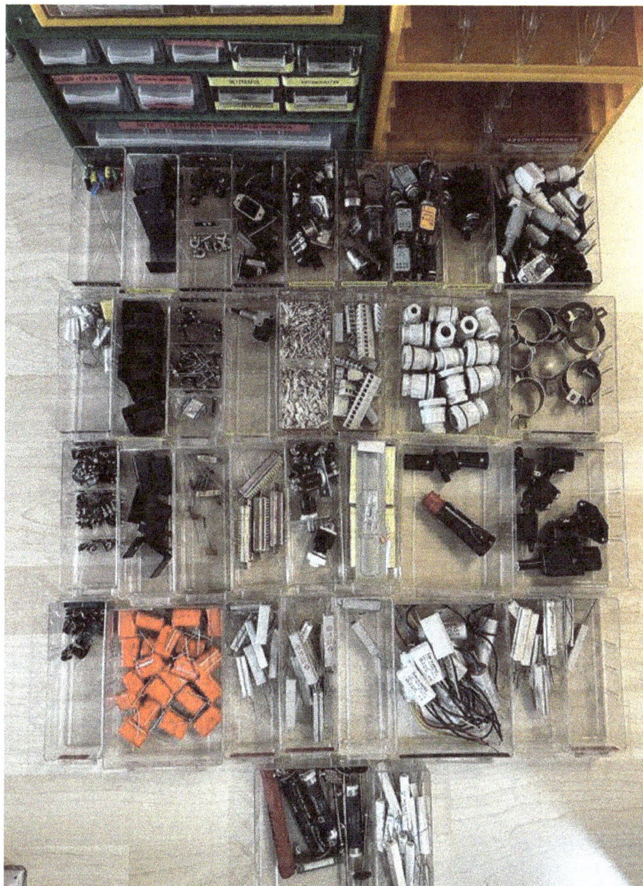

Begin by separating materials into broad categories. Electrical relays, contactors, switches, connectors, circuit breakers, industrial controls, telecommunications equipment, and miscellaneous electronic components should each have dedicated storage areas whenever practical. This organization allows you to compare similar items while gradually becoming more familiar with the characteristics of each category.

Testing can also play an important role when uncertainty exists. Commercial silver testing products are available that may help identify certain materials, particularly when used alongside visual inspection and manufacturer information. However, testing should be viewed as one part of the identification process rather than the only method. Experienced recyclers often combine visual clues, equipment type, manufacturer specifications, and testing results before deciding how to classify a component.

Keeping simple notes can also prove valuable. Recording where a particular component came from, how it was identified, and how buyers later classified it creates a personal reference library that becomes increasingly useful over time. Many experienced recyclers rely on years of observations accumulated through careful record keeping rather than memory alone.

Good organization also protects your inventory. Clean, dry storage containers help prevent unnecessary contamination while making transportation easier when it's time to sell. As your collection grows, clearly labeled containers reduce confusion and improve efficiency during every stage of the recycling process.

Tools Every Beginner Should Know

Beginning precious metals recycling does not require an elaborate workshop or expensive laboratory equipment. Most beginners can accomplish a great deal with a modest collection of hand tools, safety equipment, and organizational supplies while focusing primarily on identifying and collecting silver-bearing materials.

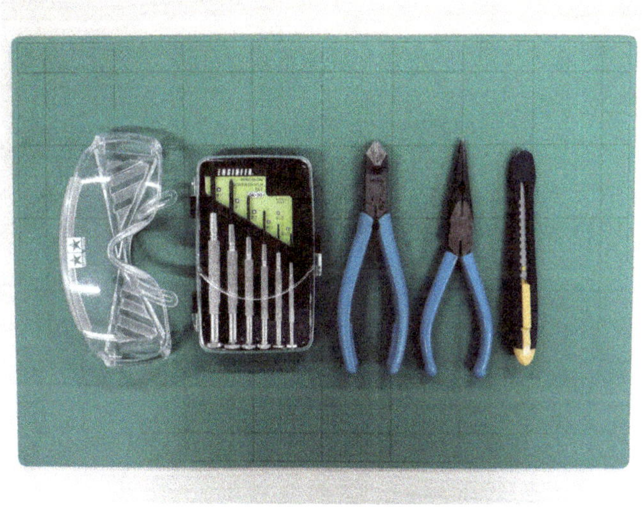

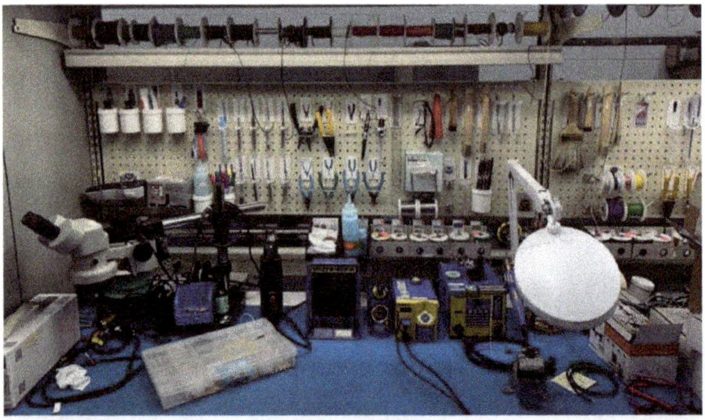

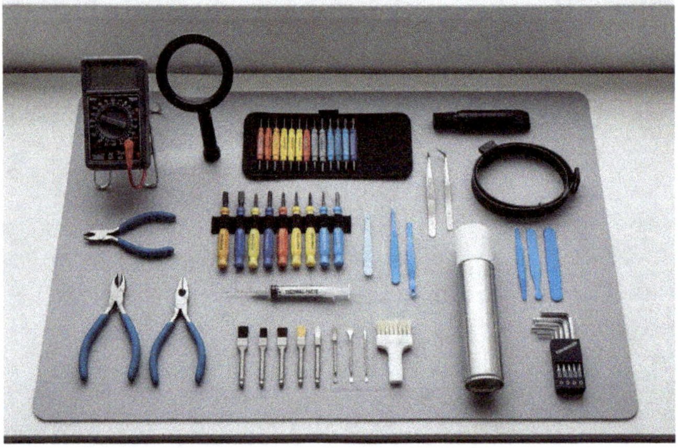

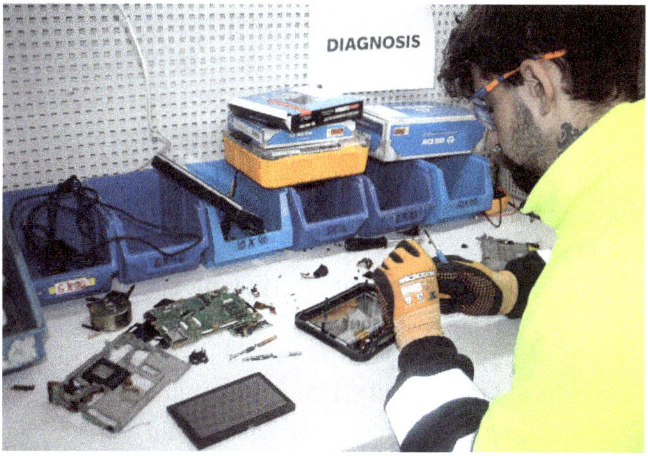
DIAGNOSIS

A quality screwdriver set is essential because electronic equipment uses a wide variety of fasteners. Phillips, flathead, Torx, and hex drivers are commonly needed when opening computers, industrial equipment, electrical devices, and household appliances. Needle-nose pliers, side cutters, tweezers, and small pry tools also make disassembly safer and more efficient.

A magnifying glass or illuminated inspection loupe can help examine contact surfaces, manufacturer markings, and fine electronic details that are difficult to see with the naked eye. A digital scale is another useful tool for estimating quantities and maintaining inventory records, while clearly labeled storage containers help keep different categories of materials separated.

Personal protective equipment should always be considered part of every beginner's toolkit. Safety glasses, durable gloves, and a clean, organized work area reduce the risk of injury when handling electronic scrap. Careful handling is especially important when working with older electrical equipment that may contain sharp edges, fragile components, or materials requiring special disposal.

Although some hobbyists become interested in chemical testing or refining equipment, these specialized tools should never replace a solid understanding of identification, organization, and responsible sourcing. In fact, many successful recyclers spend years collecting and selling silver-bearing materials without performing any chemical recovery themselves. Understanding the value of a component often proves far more important than attempting to process it.

Last Thoughts

Identifying, collecting, and preparing silver-bearing scrap is a skill that improves steadily with experience. Learning to recognize valuable components, organizing them carefully, using appropriate testing methods, and relying on basic hand tools creates a strong foundation for long-term success. Every relay, switch, connector, or industrial control you evaluate adds to your understanding of the recycling industry. In the next chapter, you'll explore what happens after silver-bearing materials have been collected, including how recovery methods work at a high level, when selling may be a better option than refining, and the practical strategies experienced recyclers use to maximize the value of their materials.

Chapter 4: Recovering and Selling Silver

After identifying and collecting silver-bearing materials, you'll eventually face an important decision: should you recover the silver yourself or sell the material to a professional buyer? This question has become increasingly common as precious metals recycling continues to grow in popularity online. While many videos make silver recovery appear simple, the reality is that recovering silver involves much more than placing components into a chemical solution. Successful recovery begins with proper identification, careful preparation, an understanding of the materials involved, and a realistic appreciation of the time, cost, and safety responsibilities associated with refining. This chapter explains how silver recovery works at a high level while helping you understand why many experienced recyclers choose different approaches depending on their goals.

Understanding Silver Recovery Methods

Before silver can be recovered, recyclers must first understand exactly what they are working with. Most silver-bearing electronic components are not solid pieces of pure silver. Instead, silver is commonly found as small contact buttons, thin plated surfaces, or specialized alloys designed to improve electrical performance. These components are usually attached to brass, copper, steel, or other base metals that must be separated before silver can be refined.

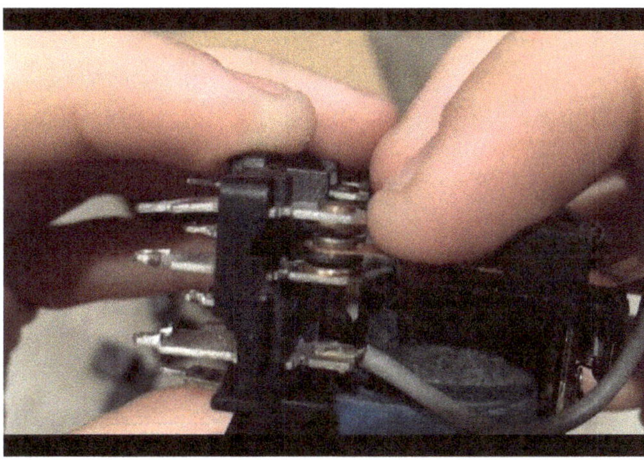

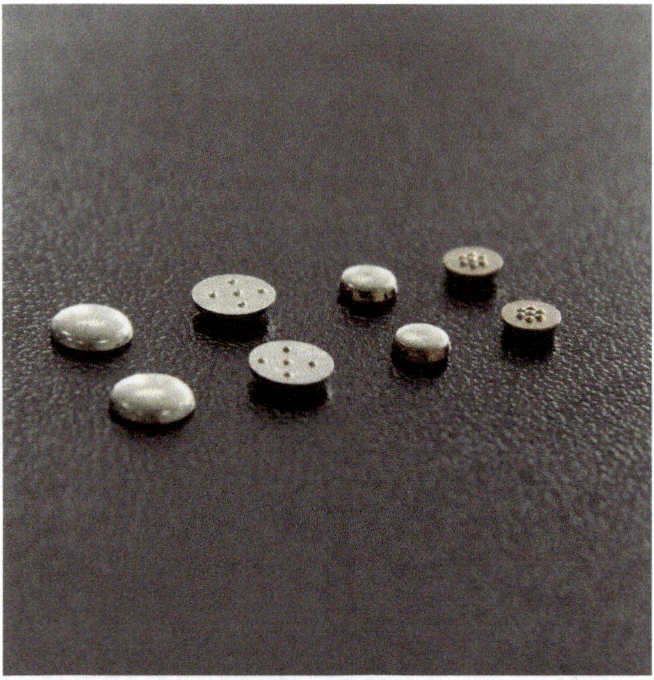

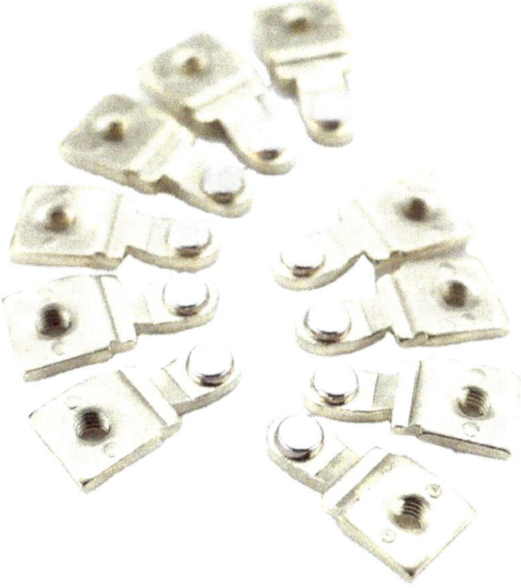

The first stage of recovery is often **mechanical separation**. Rather than immediately using chemicals, recyclers remove silver-bearing contacts from relays, switches, breakers, contactors, and other electrical equipment. Separating these components reduces unnecessary material and allows similar items to be grouped together before any additional processing begins. Many recyclers stop at this stage and simply sell the separated silver-bearing contacts to specialized refiners.

Those who continue into refining eventually encounter various chemical recovery methods. One approach discussed within the recycling community is commonly called the **acid peroxide** method. In simple terms, this process is designed to dissolve certain base metals, such as copper and brass, while leaving much of the silver behind for later recovery. Because the reaction targets the surrounding base metals first, it can gradually expose the silver contacts that remain after the supporting material has dissolved.

Another well-known refining solution is **aqua regia**, a powerful mixture used by professional refiners to dissolve precious metals so they can later be separated and purified. Unlike acid peroxide, aqua regia is capable of dissolving silver, gold, and other precious metals under controlled conditions. Once dissolved, refiners use additional chemical processes to recover the precious metals from solution. While highly effective in professional settings, aqua regia requires specialized knowledge, proper ventilation, chemical handling experience, and carefully controlled waste management procedures.

It is also important to recognize that not every silver-colored contact behaves the same way during refining. Many electrical contacts are manufactured using silver alloys containing metals such as tungsten or other materials selected for durability and resistance to electrical arcing. These alloys often require additional refining steps because the silver is chemically bonded with other metals rather than existing as pure silver. As a result, refining becomes much more complicated than simply melting the contacts into a silver bar.

Understanding these recovery methods helps explain why professional refiners possess specialized facilities and extensive technical knowledge. The purpose of this discussion is to help you understand how silver recovery works within the recycling industry, not to provide laboratory procedures for performing hazardous chemical refining.

Selling Silver-Bearing Scrap Versus Refining It

After learning how silver recovery works, many beginners are surprised to discover that recovering silver themselves is not always the most practical option. Professional precious metal refiners already possess the equipment, safety systems, environmental controls, and technical expertise required to recover silver efficiently from large quantities of material.

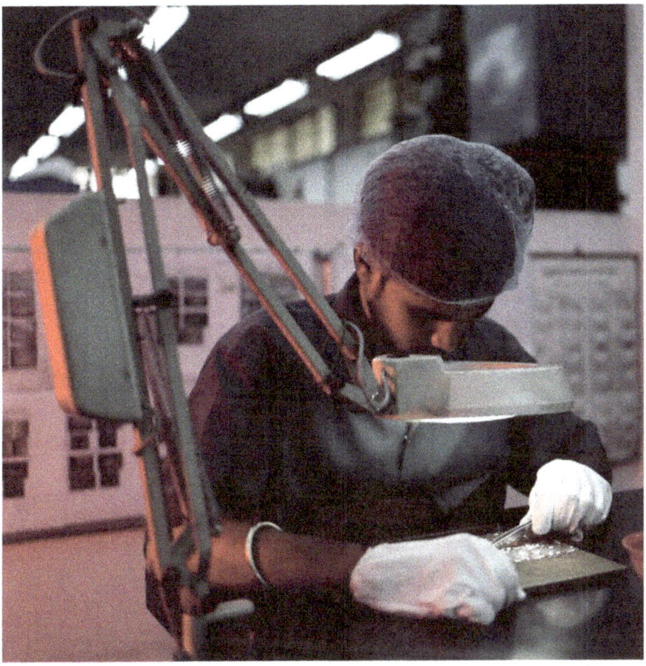

Selling silver-bearing contacts, relays, connectors, and other electrical components allows recyclers to focus on sourcing valuable materials rather than operating a chemical refining process. Buyers evaluate materials according to their composition, estimated silver content, cleanliness, weight, and the amount of processing required before refining can begin. Well-organized shipments that separate similar materials are generally easier to evaluate and may receive more accurate pricing than mixed containers of miscellaneous electronic scrap.

Refining at home also involves significant responsibilities beyond simply recovering the metal. Chemical solutions eventually become contaminated with dissolved metals and must be disposed of properly through approved hazardous waste programs. Improper disposal can create environmental hazards while violating local regulations. Appropriate personal protective equipment, chemical-resistant containers, ventilation systems, and emergency planning are all essential parts of any legitimate refining operation.

Time should also be considered. Recovering silver from small quantities of electronic scrap often requires careful preparation, multiple processing stages, patience, and ongoing monitoring. Many recyclers conclude that their time is better spent finding additional silver-bearing materials rather than operating a small refining setup. This is one of the primary reasons many successful recyclers specialize in sourcing and selling silver-bearing components while leaving the refining process to companies that perform it on an industrial scale.

Ultimately, the right choice depends on your interests, available resources, experience, and long-term goals. Some people enjoy studying refining chemistry as part of the hobby, while others prefer to focus on collecting and marketing valuable materials. Both approaches contribute to the broader precious metals recycling industry.

Maximizing Your Profits

Regardless of whether you eventually sell silver-bearing materials or pursue additional refining education, maximizing profitability begins with good decision-making long before any transaction takes place.

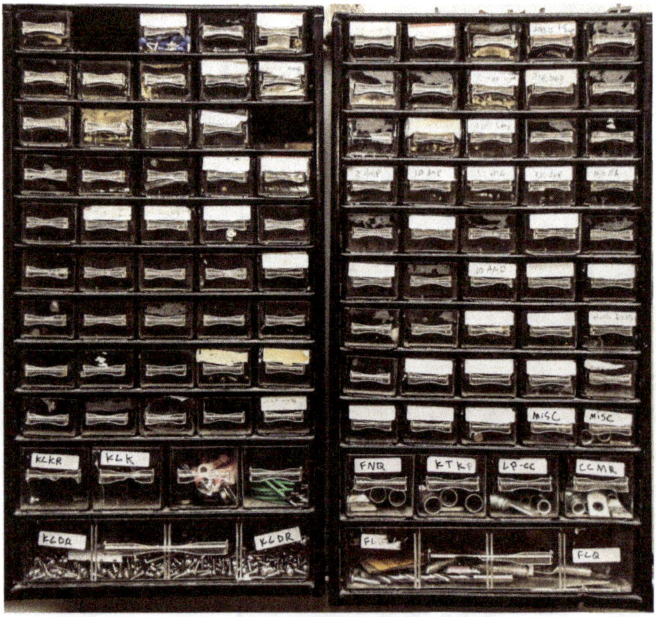
KLKR
KLK
MISC
MISC
FNQ
LP-CC
CCMR
KLDR
KLDR
FLQ

One of the simplest ways to improve returns is by carefully sorting materials before offering them for sale. Separating relays, contactors, switches, electrical contacts, connectors, and other categories allows buyers to evaluate each group more accurately while reducing unnecessary processing time.

Maintaining accurate records also becomes increasingly valuable. Recording where materials were obtained, what was paid, how buyers classified them, and which sources consistently produced worthwhile results allows you to identify trends that improve future purchasing decisions. Over time, this information becomes one of your most valuable business resources.

Comparing buyers is another important practice. Different refiners and precious metal recyclers specialize in different categories of material, use different grading systems, and may offer significantly different prices depending on current market conditions. Developing relationships with several reputable buyers helps you better understand your local market while ensuring your materials receive fair evaluation.

Finally, continue expanding your knowledge. The more familiar you become with silver-bearing components, industrial equipment, and precious metal markets, the easier it becomes to recognize opportunities that others overlook. Experience gradually improves every stage of the process, from sourcing and identification to selling and long-term business planning.

Last Thoughts

Recovering silver is about much more than chemistry. It begins with understanding the materials themselves, continues through careful identification and mechanical preparation, and ultimately leads to an informed decision about whether refining or selling best supports your goals. By learning how professional recovery methods work, appreciating the challenges involved, and recognizing the value of organized sourcing and responsible selling, you'll be well prepared to participate in the precious metals recycling industry safely, ethically, and confidently. In the next chapter, you'll learn how to transform these skills into an ethical silver recovery side business by developing reliable sources, avoiding common beginner mistakes, and building a reputation that supports long-term success.

Chapter 5: Building an Ethical Silver Recovery Side Business

Many people begin collecting silver-bearing materials simply because they enjoy recycling or are curious about precious metals. Over time, however, they often discover that consistent sourcing, careful organization, and strong business relationships can transform a hobby into a rewarding side business. Like any successful business, long-term growth depends far more on professionalism and continuous learning than on finding a single valuable item. By developing dependable sourcing channels, avoiding common mistakes, and building a reputation for honesty, you create opportunities that continue expanding as your knowledge grows.

Developing Reliable Sources

The most successful silver recyclers rarely depend on random discoveries. Instead, they build relationships with businesses and organizations that regularly replace electrical equipment, industrial controls, electronics, and other devices containing silver-bearing components.

480 VOLTS

Electrical contractors are often among the best long-term sources because they routinely replace circuit breakers, disconnect switches, contactors, motor starters, industrial relays, and electrical panels during renovations and equipment upgrades. These projects frequently generate components containing silver contacts that are no longer needed once new equipment has been installed.

Industrial maintenance companies, manufacturing facilities, commercial buildings, hospitals, schools, and utility providers also retire electrical equipment on a regular basis. Introducing yourself professionally and explaining that you responsibly recycle electrical and electronic components may create opportunities to acquire materials that would otherwise be discarded.

Computer repair businesses, electronics recyclers, and industrial surplus dealers can become valuable partners as well. Although their primary focus may not be silver recovery, they regularly encounter equipment containing silver-bearing contacts, connectors, and industrial electronics. Building dependable relationships with these businesses often produces more consistent results than searching for materials one item at a time.

Professionalism remains one of your greatest assets. Responding promptly, arriving when promised, handling materials responsibly, and communicating honestly help establish trust. Over time, suppliers often begin contacting dependable recyclers whenever additional equipment becomes available.

Avoiding Common Beginner Mistakes

Every recycler gains experience by making mistakes, but recognizing common challenges early can save considerable time, money, and frustration. One of the most common misconceptions is assuming that every silver-colored component actually contains silver. Appearance alone rarely provides enough information to determine whether a component is solid silver, silver-plated, or manufactured from an entirely different material.

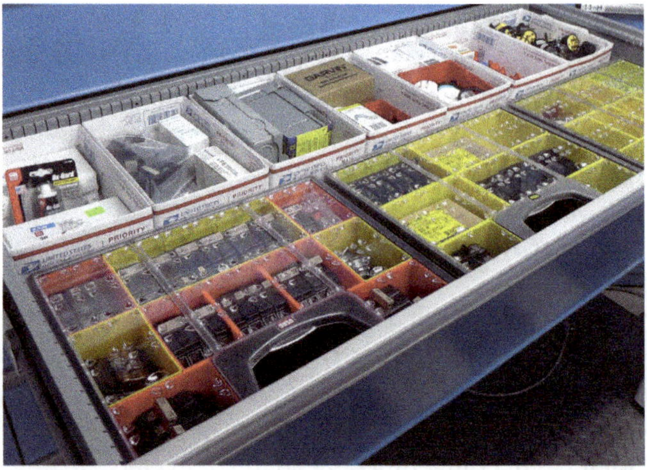

Another common mistake is collecting every electrical device without understanding where silver is typically found. While electronic equipment certainly contains valuable materials, experienced recyclers quickly learn which categories consistently produce worthwhile results and which require significant effort for relatively little return. Developing this knowledge allows you to spend your time more efficiently while avoiding unnecessary transportation and storage costs.

Some beginners also become overly focused on chemical recovery before fully understanding the materials they have collected. As discussed in the previous chapter, refining is only one part of the precious metals recycling industry and requires specialized knowledge, proper safety precautions, and responsible environmental practices. Many successful recyclers build profitable businesses by sourcing and selling silver-bearing materials without ever operating a chemical refining process themselves.

Poor organization is another obstacle that limits growth. Mixing relays, breakers, contactors, switches, connectors, and miscellaneous electronics into one container makes grading more difficult and often reduces efficiency when preparing materials for sale. Developing organized storage habits early allows your business to expand without creating unnecessary confusion later.

Finally, avoid chasing unrealistic expectations. Online videos sometimes highlight unusually valuable discoveries while overlooking the consistent effort, knowledge, and relationship-building that support long-term success. Sustainable growth almost always comes from steady improvement rather than dramatic individual finds.

Scaling Your Precious Metals Recycling Business

As your knowledge and inventory increase, you may decide to expand your operation beyond occasional recycling projects. Growth should be approached thoughtfully, allowing your systems and experience to develop alongside your inventory rather than expanding faster than you can effectively manage.

Copper
Silver Filled
Sterling
Brass
German Silver
Gold Filled

Organization becomes increasingly important as your business grows. Clearly labeled storage areas, inventory records, dedicated workspaces, and consistent sorting procedures help maintain efficiency while reducing unnecessary handling. Well-organized materials also make it easier to communicate with buyers and evaluate purchasing opportunities.

Diversifying your sourcing methods creates additional stability. Instead of depending on a single supplier or auction platform, experienced recyclers often develop relationships with electrical contractors, industrial maintenance companies, surplus dealers, government auctions, electronics recyclers, computer repair shops, manufacturers, and liquidation firms. Multiple sourcing channels help maintain a consistent flow of materials throughout the year.

Financial discipline also contributes to long-term success. Recording purchases, transportation costs, sales, buyer information, and inventory levels provides valuable insight into which activities generate the strongest returns. These records help identify profitable trends while supporting smarter decisions as your business expands.

Most importantly, continue investing in your knowledge. New technologies, changing electrical equipment, evolving recycling markets, and updated environmental regulations continually shape the precious metals recycling industry. Recyclers who remain curious and continue learning are often the first to recognize emerging opportunities while adapting successfully to industry changes.

Your reputation ultimately becomes one of your greatest competitive advantages. Suppliers appreciate dependable recyclers who operate ethically, buyers value honestly represented materials, and long-term business relationships are built on trust rather than short-term profit. Protecting that reputation is one of the best investments you can make.

Last Thoughts

Building an ethical silver recovery side business is a gradual process based on knowledge, consistency, and integrity. Every successful recycler begins by learning where valuable materials are found, developing dependable sourcing relationships, and improving their skills one project at a time. By avoiding common beginner mistakes, organizing your operation carefully, and focusing on long-term relationships rather than quick profits, you'll create a strong foundation for sustainable growth. In the next chapter, you'll explore advanced opportunities in precious metal recycling, discover valuable metals beyond silver, and gain a broader understanding of how responsible recycling supports the circular economy while creating new opportunities for the future.

Chapter 6: Advanced Opportunities in Precious Metal Recycling

As your experience with silver-bearing materials grows, you'll begin recognizing opportunities that extend far beyond individual relays, switches, and electrical contacts. The precious metals recycling industry includes a wide variety of valuable materials found throughout electronics, industrial equipment, telecommunications systems, renewable energy technologies, and manufacturing operations. Successful recyclers understand that every piece of equipment should be evaluated as a collection of valuable resources rather than a single item. By expanding your knowledge and continuing to learn, you'll be prepared to recognize new opportunities as technology continues to evolve.

Beyond Silver: Gold, Palladium, and Other Valuable Metals

Silver is only one of several precious metals commonly recovered through electronic recycling. Manufacturers also use gold, palladium, platinum, and other specialized metals in applications where corrosion resistance, electrical performance, or long-term durability are especially important.

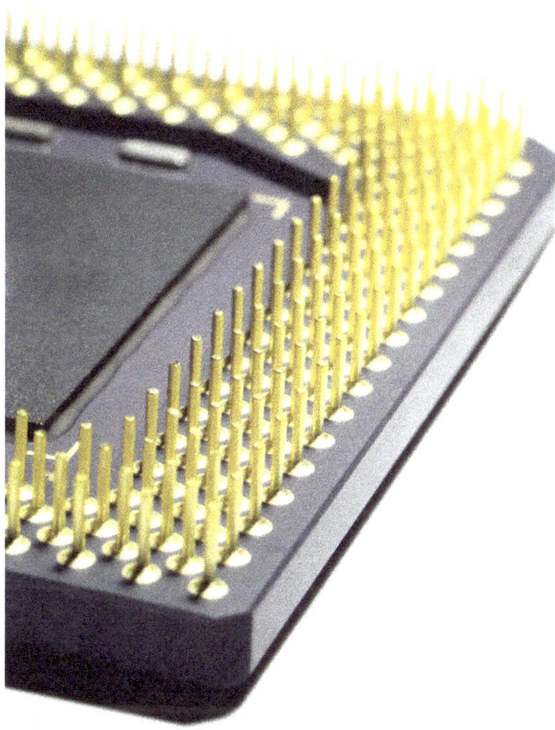

1R0

Gold is frequently found on connector pins, edge connectors, processors, memory modules, and telecommunications equipment because it resists corrosion while maintaining excellent electrical conductivity. Palladium appears in certain electronic components, multilayer ceramic capacitors, and specialized industrial equipment. Platinum is less common but may be found in laboratory equipment, industrial applications, and catalytic systems.

Many electronic devices contain several valuable materials at the same time. A server, for example, may include copper wiring, aluminum heat sinks, stainless steel hardware, circuit boards containing gold-plated connectors, silver-bearing contacts, and electric motors. Evaluating equipment from this broader perspective allows recyclers to recover value from multiple material streams rather than focusing exclusively on one precious metal.

The same principle applies to industrial electrical equipment. A single motor starter, relay cabinet, or control panel may contain silver contacts, copper conductors, aluminum bus bars, steel enclosures, and electronic control boards. Understanding how these materials work together helps you make more informed decisions when purchasing, sorting, and selling electronic scrap.

As technology changes, new products continue entering the recycling stream. Electric vehicles, renewable energy systems, battery storage equipment, telecommunications infrastructure, and advanced manufacturing equipment all create opportunities for recyclers willing to continue expanding their knowledge.

Understanding the Circular Economy

Precious metals recycling is an important part of the **circular economy**, a system designed to keep valuable materials in productive use for as long as possible. Rather than continuously extracting new raw materials from the earth, recycling allows manufacturers to recover metals already present in existing products and return them to future manufacturing.

Every year, millions of computers, industrial machines, electrical systems, communication devices, medical instruments, and consumer electronics reach the end of their useful lives. Many of these products still contain valuable metals that can be recovered through responsible recycling. Recovering these materials reduces waste, conserves natural resources, and decreases the environmental impact associated with mining and processing new raw materials.

The circular economy also emphasizes extending the useful life of products whenever possible. Some electronic equipment can be repaired, refurbished, upgraded, or reused before eventually being recycled. Responsible recyclers recognize that reuse is often preferable to dismantling equipment that still serves a useful purpose.

This broader perspective highlights the important role recyclers play within modern manufacturing. Every relay, connector, circuit board, electrical contact, and industrial control recovered responsibly contributes to a larger system that supports resource conservation while helping manufacturers obtain valuable raw materials for future products.

Understanding this connection helps reinforce why ethical sourcing and responsible recycling practices are so important. The work performed by recyclers extends well beyond individual profits. It supports industries, communities, and environmental stewardship around the world.

Long-Term Success in Precious Metal Recycling

Long-term success in precious metals recycling is built through steady learning rather than short-term discoveries. Markets change, technology evolves, and new equipment enters the recycling stream every year. Recyclers who remain curious and adaptable consistently place themselves in stronger positions to recognize opportunities before they become widely known.

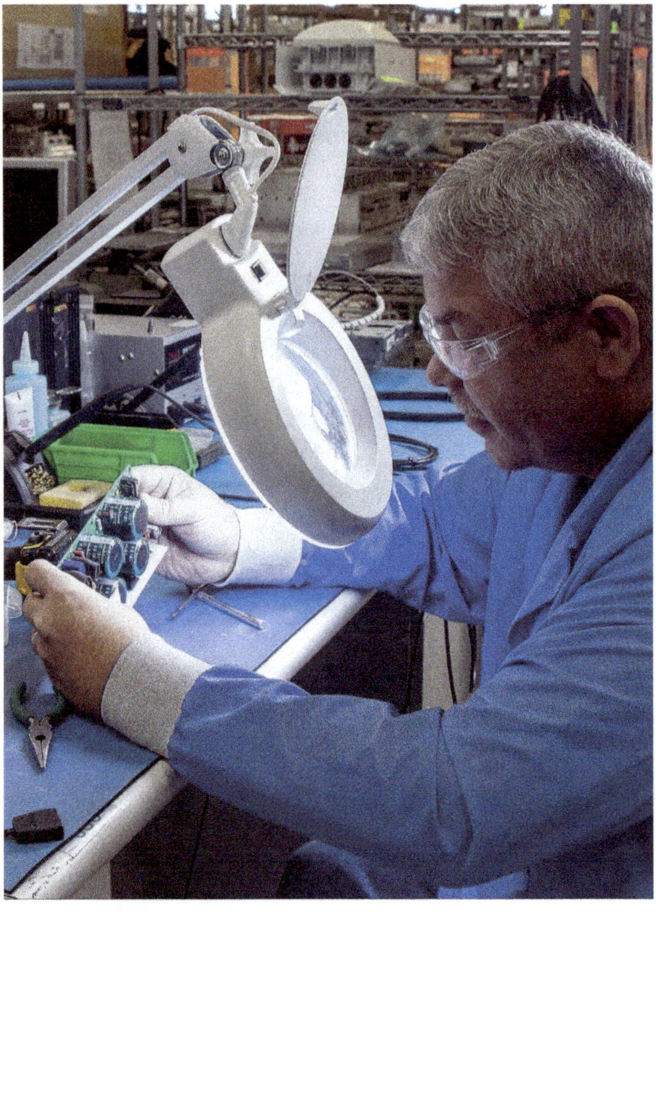

IT EQUIPMENT
BATTERIES
SMALL APPLIANCES

Building strong professional relationships continues to be one of the most valuable investments you can make. Electrical contractors, industrial maintenance companies, computer repair businesses, electronics recyclers, auction companies, manufacturers, surplus dealers, and professional buyers all contribute to the precious metals recycling ecosystem. Treating every interaction with honesty and professionalism creates trust that often leads to future opportunities.

Maintaining organized records also becomes increasingly valuable as your business grows. Tracking purchases, inventory, transportation costs, sales, buyer preferences, and sourcing locations helps identify trends that improve future decision-making. Small improvements made consistently often produce greater long-term results than dramatic individual transactions.

Continuing your education is equally important. Reading industry publications, attending surplus auctions, visiting recycling facilities, learning about emerging technologies, and studying how manufacturers use precious metals all contribute to your professional development. The recycling industry is constantly evolving, and those who continue learning are often the first to recognize new opportunities.

Above all, protect your reputation. Markets may fluctuate and technologies may change, but integrity remains one of the few competitive advantages that never loses its value. Honest communication, ethical sourcing, responsible environmental practices, and respect for both suppliers and buyers create a foundation that supports lasting success regardless of market conditions.

Last Thoughts

Precious metals recycling extends far beyond recovering silver from individual components. It is part of a much larger industry that supports manufacturing, sustainability, resource conservation, and responsible waste management around the world. By expanding your knowledge beyond silver, understanding the importance of the circular economy, and committing yourself to continuous learning, you'll be well prepared to grow alongside this evolving industry. Every electrical contact, industrial relay, electronic connector, and silver-bearing component represents more than recyclable material. It represents an opportunity to recover valuable resources, reduce waste, and build a rewarding business founded on knowledge, integrity, and responsible recycling.

Image Sources, References, and Additional Resources

Public Domain Image Sources

The public domain images used throughout this book were obtained or adapted from the following trusted repositories. These collections provide educational, historical, industrial, and government-produced images that are generally suitable for educational publishing. Always verify the licensing status of individual images before publication.

- Library of Congress Prints & Photographs Division
  https://www.loc.gov/pictures/

- U.S. National Archives
  https://www.archives.gov/research

- U.S. Department of Energy Image Gallery
  https://www.energy.gov/

- U.S. Geological Survey (USGS) Multimedia Gallery
  https://www.usgs.gov/media/images

- National Institute of Standards and Technology (NIST) Image Gallery https://www.nist.gov/
- U.S. Environmental Protection Agency (EPA) Multimedia Resources https://www.epa.gov/
- NASA Image and Video Library https://images.nasa.gov/
- Wikimedia Commons (Public Domain and Public Domain Marked Images) https://commons.wikimedia.org/
- Rawpixel Public Domain Collection https://www.rawpixel.com/category/53/public-domain
- Openverse https://openverse.org/

References

Basel Convention. *Technical Guidelines on the Environmentally Sound Management of Waste Electrical and Electronic Equipment (WEEE).*

Institute of Scrap Recycling Industries (ISRI). *Electronics Recycling Resources.*

Sustainable Electronics Recycling International (SERI). *Responsible Electronics Recycling Standards*.

U.S. Environmental Protection Agency (EPA). *Electronics Donation and Recycling*.

U.S. Geological Survey (USGS). *Mineral Commodity Summaries*.

National Institute of Standards and Technology (NIST). *Materials Science and Engineering Resources*.

Occupational Safety and Health Administration (OSHA). *Chemical Safety and Hazard Communication*.

National Institute for Occupational Safety and Health (NIOSH). *Chemical Hazards and Laboratory Safety*.

United Nations Institute for Training and Research (UNITAR). *Global E-Waste Monitor*.

International Telecommunication Union (ITU). *Global E-Waste Statistics*.

Additional Resources

Institute of Scrap Recycling Industries (ISRI)
https://www.isri.org/

Sustainable Electronics Recycling International (SERI)
https://sustainableelectronics.org/

U.S. Environmental Protection Agency, Electronics Donation and Recycling
https://www.epa.gov/recycle/electronics-donation-and-recycling

U.S. Geological Survey (USGS)
https://www.usgs.gov/

National Institute of Standards and Technology (NIST)
https://www.nist.gov/

Occupational Safety and Health Administration (OSHA)
https://www.osha.gov/

National Institute for Occupational Safety and Health (NIOSH)
https://www.cdc.gov/niosh/

Global E-Waste Statistics Partnership
https://globalewaste.org/

The Silver Institute
https://www.silverinstitute.org/

London Bullion Market Association (LBMA)
https://www.lbma.org.uk/

About the Author

Roggie "Raj" Clark is an entrepreneur and a 10-year marketing professional with a proven track record in e-commerce and digital business growth. Having built and managed successful online stores with thousands of sales and consulted for numerous marketing clients, Roggie has mastered the intricacies of online marketing, branding, and consumer behavior. He has also written several popular books on digital marketing, SEO (search engine optimization), and programming (HTML, CSS, and JavaScript).

As a former marketing manager at a fintech company—a role he secured through his own marketing and financial expertise—Roggie gained firsthand experience in financial strategy and wealth-building. This journey inspired him to share insights with amateur investors, helping them save money, make smarter financial decisions, and build long-term wealth. Raj currently lives in Maryland with his sister and niece.